WHAT GROWS IN MY

GREENHOUSE

LINDA BLACKMOOR

ISBN: 978-1-966417-14-9 (PRINT)

PUBLISHED BY QUILL PRESS. LINDA BLACKMOOR'S TITLES MAY BE PURCHASED IN BULK FOR EDUCATIONAL, BUSINESS, FUNDRAISING, OR SALES PROMOTIONAL USE. FOR INFORMATION, PLEASE EMAIL HELLO@LINDABLACKMOOR.COM

FIRST PRINT EDITION: 2025

LINDA BLACKMOOR
WWW.LINDABLACKMOOR.COM

POTHOS

(EPIPREMNUM AUREUM)

Pothos is a charming, trailing vine with heart-shaped leaves dappled with golden hues. Native to the tropical forests of the Solomon Islands, it brings a touch of exotic wonder into any home.

- Care Tip: Give it bright, indirect light and water when the top inch of soil feels dry.

- Did You Know? Pothos is also known as "devil's ivy" because it can thrive in many conditions, making it a resilient little explorer!

SPIDER PLANT

(CHLOROPHYTUM COMOSUM)

The Spider Plant sways gracefully with slender, arching leaves and tiny offshoots, called spiderettes, that dangle like playful ornaments. Native to South Africa, it has been a friendly green companion in homes for generations.

- Care Tip: Place it in bright, indirect light and water when the soil is slightly dry—avoiding soggy roots keeps it happy.

- Did You Know? Those little spiderettes can grow into new plants, making Spider Plant a living family tree!

SNAKE PLANT

(SANSEVIERIA TRIFASCIATA)

With its tall, sword-like leaves, the Snake Plant stands proudly like a green sentinel. Hailing from the arid regions of West Africa, it's built to thrive even when care is a bit forgetful.

- Care Tip: It loves low to bright, indirect light and only needs water when the soil is dry—be careful not to overwater!

- Did You Know? Snake Plants are celebrated for their ability to clean indoor air, acting like a natural little purifier.

PEACE LILY

(SPATHIPHYLLUM)

The Peace Lily enchants with glossy, dark green leaves and elegant white blooms that seem to whisper serene secrets. Native to the tropical Americas, it graces any space with quiet beauty and calm.

- Care Tip: Enjoys bright, indirect light and a steady drizzle of water to keep its soil evenly moist.

- Did You Know? Despite its name, the Peace Lily isn't a true lily—it's a gentle ambassador of tranquility in the plant kingdom.

DRACAENA

(DRACAENA SPP.)

Dracaena sways with long, slender leaves that create a natural silhouette, as if dancing in soft light. With roots in the tropical and subtropical regions of Africa and Asia, it brings a whisper of faraway lands into your room.

- Care Tip: Place it in bright, indirect sunlight and water sparingly, letting the topsoil dry out before the next drink.

- Did You Know? Dracaena is a natural air cleaner, helping to keep your indoor space fresh and inviting!

FIDDLE LEAF FIG

(FICUS LYRATA)

The Fiddle Leaf Fig captivates with large, violin-shaped leaves that seem to play a silent symphony of nature. Native to the lush forests of West Africa, it adds a dramatic, musical flair to any room.

- Care Tip: Keep it in bright, filtered light and water moderately—allow the top soil to dry out before watering again.

- Did You Know? The Fiddle Leaf Fig gets its name from the distinctive shape of its leaves, which resemble a fiddle.

RUBBER PLANT

(FICUS LYRATA)

The Rubber Plant dazzles with thick, glossy leaves that shine like little emeralds. Originating from the rainforests of South and Southeast Asia, it carries a legacy of tropical charm and robust growth.

- Care Tip: Enjoy bright, indirect light and water when the top inch of soil is dry to keep its leaves lustrous.

- Did You Know? Rubber Plant's latex was once a vital source of natural rubber, used long before synthetic alternatives were invented!

ZZ PLANT

(ZAMIOCULCAS ZAMIIFOLIA)

The ZZ Plant boasts shiny, waxy leaves that catch the light even in the dimmest corners. Native to Eastern Africa, it's a marvel of resilience, thriving with very little fuss.

- Care Tip: It flourishes in low light and only needs water when the soil has completely dried out.

- Did You Know? ZZ Plants store water in their thick rhizomes, making them remarkably drought-tolerant!

PHILODENDRON

(PHILODENDRON SPP.)

With its lush, heart-shaped leaves, the Philodendron creates a warm, inviting atmosphere that whispers tropical secrets. Hailing from the vibrant rainforests of Central and South America, every leaf tells a story of nature's love.

- Care Tip: Provide moderate to bright, indirect light and water when the top layer of soil feels dry.

- Did You Know? Some Philodendron species naturally climb by using their leaves to gently cling to surfaces, showcasing nature's ingenious designs!

ENGLISH IVY

(HEDERA HELIX)

English Ivy drapes its winding vines with delicate, evergreen leaves that evoke images of ancient castles and secret gardens. Native to Europe, Asia, and North Africa, it's a timeless traveler that adds elegance to any wall or window frame.

- Care Tip: It loves bright, indirect light and benefits from regular watering to keep the soil evenly moist.

- Did You Know? English Ivy has been celebrated in folklore and art for centuries, often symbolizing fidelity and eternal life!

CHINESE EVERGREEN

(AGLAONEMA)

The Chinese Evergreen enchants with patterned, glossy leaves in varying shades of green, sometimes touched with hints of silver. From the tropical regions of Asia and New Guinea, it brings a whisper of the exotic into everyday life.

- Care Tip: Thrives in low to medium light and prefers a moderate watering schedule to keep its soil lightly moist.

- Did You Know? Chinese Evergreen is prized for its air-purifying qualities, helping to create a fresher indoor atmosphere!

ALOE VERA

(ALOE BARBADENSIS MILLER)

Aloe Vera stands as a spiky, succulent wonder, with thick leaves hiding a soothing, healing gel within. Native to the warm, sunny Arabian Peninsula, it has been treasured for its natural remedies for centuries.

- Care Tip: Give it bright, indirect light and water sparingly—allow the soil to dry out completely between waterings.

- Did You Know? The gel inside Aloe Vera leaves is famous for soothing minor burns and skin irritations, earning it a place in natural medicine cabinets!

JADE PLANT

(CRASSULA OVATA)

The Jade Plant sparkles with plump, coin-like leaves that exude a sense of good luck and prosperity. Native to South Africa, this charming succulent has long been a symbol of positive energy in homes around the world.

- Care Tip: It loves bright light and only needs water when the soil is completely dry—overwatering can upset its balance!

- Did You Know? Jade Plants are often called "money trees" because many believe they attract wealth and abundance!

BROMELIAD

(VARIOUS WITHIN BROMELIACEAE)

Bromeliads burst with vibrant, tropical colors and unique shapes, often forming a central cup that collects water like a mini rainforest pond. Native to the tropical Americas, they bring a splash of exotic flair into your space.

- Care Tip: Provide bright, indirect light and water both the soil and the central cup regularly to mimic their natural habitat.

- Did You Know? The water-holding cup of a Bromeliad can become a tiny ecosystem, sometimes hosting little critters and even miniature plants!

BOSTON FERN

(NEPHROLEPIS EXALTATA)

The Boston Fern unfurls its delicate, feathery fronds like a cascading waterfall of green, evoking the mystery of an enchanted forest. Native to tropical regions, it transforms any space into a cool, refreshing woodland retreat.

- Care Tip: Thrives in bright, indirect light and needs consistently moist soil to keep its fronds lush and vibrant.

- Did You Know? Boston Ferns are natural air purifiers, helping to refresh your indoor space like a gentle forest breeze!

www.ingramcontent.com/pod-product-compliance
Lightning Source LLC
Chambersburg PA
CBHW060834270326
41933CB00002B/88